A GUIDE ENGINEERING AND ARCHITECTURAL DESIGN SERVICES CONTRACTS

WHAT EVERY PROJECT MANAGER NEEDS TO KNOW

John M. Lowe, Jr., P.E.

Copyright © 2013 by John M. Lowe, Jr. All rights reserved.

Library of Congress Control Number: 2013916658

ISBN-13: 978-1492295563
ISBN-10: 1492295566

ACKNOWLEDGMENTS

Kathleen L. Lowe, my wife, who encouraged me to write this book and has been the primary editor and proofreader.

Susan W. Menkes, Esq. (1952-2005), my contract language mentor, without whose wise counsel much of this book could not have been written.

Henry W. Haeseker, P.E., my professional mentor, who taught me much of what I know about managing professional services contracts, and suggested and edited the Summary and Recommendations section of the book.

Kermit L. Prime, Jr., P.E., my friend and professional colleague, who created an opportunity for me to develop the text of this book.

E. Woody Lingo, P.E., my friend of many years and professional colleague, who contributed to the writing of the section on Subsurface Risks.

Dennis L. Barton, my friend and professional colleague whose proofreading and editing efforts are greatly appreciated.

Karen A. Tatman, P.E., my professional colleague and most skillful project manger that I ever worked with, who provided the exhibits for Issues Log, RFI Log, and Shop Drawing Log.

Charlotte A. Davidson, my friend and professional colleague, who prepared the exhibit for Change of Scope Form.

PREFACE

It seems to me that there are already a lot of books that are intended to help design professionals manage their projects, so why another book that primarily addresses how to manage the professional services contract? I found that during my half-century career as a consulting engineer, a lot of what I had to learn was as much about managing my contracts as it was about managing my projects. Unfortunately, there was little available on the subject of managing the contract. There were so many nuances in connection with managing the contract that could only be learned through what was often a painful experience. So, I decided to pass along to other design professionals much of what I have learned. Because the line between contract management and project management is not always clearly defined, there is some overlap in this book.

I am not an attorney and this book does not offer legal advice.

This book can be used in several ways:
As a primer on professional services contracts –The topics are arranged in the chronological order commonly experienced during a project. Accordingly, they can be studied without reference to a specific project in preparation for executing any project from start to finish. The book could also be used as a text to introduce college students to the basics of contracting for professional services.

As a resource during a project – As certain contracting topics come into play during a project, the Table of Contents of the book can be referred to for guidance.

As a source of topics that can be used to create a culture of professional liability issues awareness – By discussing one of the book's topics at a weekly staff meeting throughout the year, everyone involved in the design process can be sensitized to the importance of correctly dealing with issues that may otherwise result in a professional liability claim.

A GUIDE TO MANAGING ENGINEERING AND ARCHITECTURAL DESIGN SERVICES CONTRACTS

TABLE OF CONTENTS

Section	**Title**	**Page**

I. Introduction to Managing Engineering and Architectural Design Services Contracts 7
    A. Purpose 7
    B. Definitions 7
    C. Overview 8
II. Elements of the Contract 9
    A. Realistic Expectations 9
    B. Elements of the Contract and its Terminology 9
        General Terms and Conditions 10
            1. Successors and Assigns 10
            2. Ownership of Documents 10
            3. Opinion of Probable Construction Costs 10
            4. Standard of Care 11
            5. Deliverables 11
            6. Limit of Liability 12
            7. Insurance 12
            8. Payment Terms 13
            9. Electronic Communications 13
            10. Information Provided by Others 13
            11. Indemnification 14
            12. Change Management 15

13. Purchase Orders 15
14. Subsurface Risks 16
15. Third Party Claims 19
16. Stop Work Authority 20
17. Curing a Breach 22
18. Dispute Resolution 22
19. Termination/Suspension of Consultant's
   Services 23
Expectation Management 23
Scope of Work 25
Schedule 26
Budget 27
Basis of Payment 28
Payment Terms 29
Contingent Fee Contracts 29
Contract Signatory Requirements 31
Notice to Proceed (NTP) 31
Insurance Certificates 31
Contractor-Proposed Value Engineering 32

III. Implementing the Contract 33
   A. Creating a culture of professional liability issues
      awareness 33
   B. Minimize professional liability risk 33
      Sharing the Scope of Work (SOW) 33
      Scope Change Identification and Documentation 34
      Awareness of the Exclusions in the Professional
      Liability Policy 38
   C. Improve Efficiency 39
      Planning for Meetings 39
      Meeting Agenda 41
      Meeting Sign-In Sheet 42
      Meeting Summary/Minutes 46
      Permitting 49
      Documentation 50
         1. Telephone Calls 51
         2. Issue Tracking Log 52
         3. Site Photography 53

  Quality Assurance/Quality Control Plan (QA/QC
  Plan) 55
  Certifications 56
  Constructability Reviews 58
  Construction Contingency Fund 58
  Accounts Receivable 60
  Shop Drawing Management 63
 D. Liability 65
  Liability 65
  Damages 65
  Negligence 66
  Negligence Avoidance 67
  Professional Liability Claims Process 68

IV. During Bidding and Construction 71
 A. Before Construction Begins 71
 B. During Construction 73
  Request for Information (RFI) 73
  Shop Drawing Review 75
  Performance Bonds 75
  Record Documents 79

V. Summary and Recommendations 81
 A. Summary 81
 B. Recommendations 82

EXHIBITS

Exhibit	Title	Page
1	Change of Scope Form	36
2	Contract Amendment Form	37
3	Meeting Checklist	40
4	Meeting Agenda	42
5	Generic Sign-In Sheet	44
6	Frequent Attendee Sign-In Sheet	45
7	Seminar Sign-In Sheet	45
8	Meeting Summary	48
9	Issues Tracking Log	53
10	Site Photos	55
11	RFI Tracking Log	74
12	Shop Drawing Review Log	77

For information regarding obtaining the exhibits in their native format (MS Word and Excel), visit the author's website, www.loweconsultingllc.com

Managing Engineering and Architectural Design Services Contracts

I. Introduction to Managing Engineering and Architectural Design Services Contracts

A. **Purpose**. The purpose of this book is to provide the practicing engineer or architect with information that will improve the efficiency of the design and construction process, reduce the probability of a professional liability claim, and improve the likelihood of the expectations of the parties to the contract being met. The central theme of this book is achieving expectation management. Almost all of the information contained herein is intended to enhance that process. While the book is primarily directed to practitioners that serve as Consultants to Clients with whom they have a contract, the philosophies and procedures contained herein should be beneficial to Clients as well.

B. **Definitions**. The word "Consultant" is the term that is used to represent design professionals (engineers and architects) collectively. The word "Client" is the term used to represent the entity that is the Owner of the project that is being designed. The word "contract" refers to a legal document between two or more parties, one of which is a Client and one of which is a Consultant. The contract is the document in which the expectations of both the Client and the Consultant are codified in an effort to give each party a clear understanding as to what can be expected of the other party.

C. **Overview**. The management of a professional services contract begins with considerations regarding how the Consultant is selected. This is usually followed by developing and negotiating the contract, followed by specific actions that the Consultant can take during the design process. Finally, what the Consultant can do to administer the construction contract in an efficient and beneficial way is addressed.

II. Elements of the Contract

A. Realistic Expectations

From the Consultant's first contact with the Client, expectations on the part of both parties are being created. The Client is forming an expectation of what can be expected from the Consultant and vice versa. For the Consultant, it is important that, while presenting its qualifications and potential benefits to the Client in the most favorable light, this information should be realistic and not overstated. Frequently, Clients are persuaded by the marketing information to expect a higher level of performance than they should. When this situation has resulted in unmet expectations in the past, some Clients protect themselves by requiring on future contracts that the Consultant's Statement of Qualifications be an attachment to the contract for professional services. Statements such as the following should be avoided:

> "We assure you that our skill as design professionals will result in contract documents of the highest possible quality. Further, change orders during construction will be the lowest in the industry."

When included in a contract, this statement offers a level of service far in excess of that required and may even void professional liability insurance coverage. Likewise, an aspirational goal that addresses employee performance in a Code of Ethics can be problematic if it describes performance greater than that required in the Standard of Care.

B. Elements of the Contract and its Terminology

This process involves reaching agreement between the Client and the Consultant regarding the business relationships, frequently referred to as the General Terms and Conditions, and the Scope, Schedule, and Budget for a specific project. The contract establishes the basis of all other elements of contract expectation management.

<u>General Terms and Conditions</u>, frequently referred to as "Boilerplate," defines how the Client and Consultant will relate to each other in a business relationship without regard to a specific project. While not exhaustive, the following is a list of several of the critical "Boilerplate" items that if not clearly and equitably stated will likely result in misunderstandings and confrontational situations:

1. Successors and Assigns – This contract clause is normally stated as a negative that prohibits one or both parties from assigning their interests in the contract to a third party. It is important that both, not only one party, be so constrained.

2. Ownership of Documents – Generally speaking, it is preferable for the Consultant to maintain ownership, and thereby control, of its instruments of service or documents. If the Client insists on having ownership of the documents, then the contract should contain a provision in which the Client accepts full responsibility for any damages that occur as a result of the Client reusing or modifying the documents and indemnifies the Consultant against any losses resulting there from.

3. Opinion of Probable Construction Costs – This is a key element of expectation management. The level of accuracy and predictability of costs provided to the Client compared with that which will be received from bidders must be clearly explained in the contract. Otherwise, a Client may believe that the Consultant's opinion of probable construction costs, sometimes referred to as a "cost estimate", is an amount that absolutely will not be exceeded when the

bids are received. Language suitable for this provision is readily available from contract templates prepared by various professional societies. Some contracts require that if bids exceed a predetermined Client budget the Consultant must redesign the project at no additional cost to the Client to bring the costs within the budget. In this case, it is imperative that the Consultant avail itself of the opportunities during the design to notify the Client of concerns that the budget may not be adequate to obtain the project as scoped. This type of contract usually contains milestones at which the Consultant must provide an opinion of probable construction cost for consideration by the Client. If the opinion of probable construction cost exceeds the Client's budget, the Client may then either revise the scope or the budget to reconcile the difference.

4. Standard of Care – Standard of Care is the name given to a contract provision that establishes the level of perfection required in the Consultant's deliverables. Common Law, on which obtaining professional design services is based, requires that a Consultant's performance only be at that level of skill or care ordinarily demonstrated by members of the Consultant's profession. Attempts to expand this definition can create a situation for which professional liability insurance may not be obtainable. Language suitable for this provision is also readily available from contract templates prepared by various professional societies.

5. Deliverables – The terminology used to describe that which will be delivered to the Client is very important. The use of inappropriate language can nullify the protection afforded by the "Standard of Care" clause in the contract. The deliverables that are to be provided to the Client are intellectual property. They are "instruments of service". Unfortunately,

often deliverables are described in a contract as "work product". The problem with use of the term "work product" is that it can be argued that what is being provided is a "product," which it is not. If that argument is successful and the deliverables are declared products, then the realm of liability can be changed from "professional liability" to "product liability," to which the law applies a very different standard of performance. In certain forms of product liability, proving negligence is not a required element for the establishment of liability. Further, since professional liability insurance affords protection only for negligent acts, it may not be available if the deliverables are described as "work product". Accordingly, whenever possible, describing deliverables as "work product" should be avoided.

6. Limit of Liability – In many cases, fees charged for professional design services are significantly lower than the potential exposure to risk associated with construction of the project. For a more equitable situation, limiting the Consultant's liability to a fixed amount or one related to the Consultant's fee is desirable. Additionally, some professional liability carriers offer a premium discount to Consultants whose contracts contain the Limit of Liability provision.

7. Insurance – The contract should identify specific amounts of coverage for the five most common types of insurance, i.e. General Liability, Auto Liability, Worker's Compensation, Employer's Contingent Liability, and Professional Liability. The contract language regarding Professional Liability insurance must be carefully crafted to explain that, unless otherwise specified, the amount provided is the coverage that is in place for the Consultant during the policy period. This means that the amount available for any particular claim is the net of the policy face

amount after any prior claims are paid during the policy year. For the Client to actually have continuous full coverage for its project would in most cases require that Project Insurance be obtained. Insurance provisions of a contract should be provided by the Consultant to its insurance broker for review.

8. Payment Terms – How and when the Consultant is to be paid for its services should be clearly defined in the contract. The length of time available to the Client to pay the Consultant's invoice without penalty should be stipulated along with any penalty, such as interest on unpaid invoices, for invoices not paid within the allowable time.

9. Electronic Communications – Much of the communications between the Client and Consultant during the design and construction will be accomplished through electronic communications such as e-mail and Fax. The authority of electronic communications, such as e-mail, should be clearly defined in the contract. Electronic communications from the Client that result in changes to the contract scope, schedule, or budget must be followed up with a written contract amendment that addresses the change and is signed by both parties.

10. Information Provided by Others - Whether the Consultant is taking over a design started by others or the Client is to provide information that the Consultant is to use in the design, several considerations should be addressed in the contract. These include the following:

    - List the information to be provided under the "Client's Responsibilities" provision.

- Include a schedule establishing when the information is to be available.

- State to what degree Consultant can rely on the Client-provided information, i.e., with or without independent review.

- If the Client wants the Consultant to review the provided information, that activity should be a task item for which compensation is provided and for which the limit of the Consultant's responsibility is agreed upon.

- Seek agreement that the Client will provide additional new or supplemental information that is identified as being needed during the design.

- Seek agreement that the Client will indemnify the Consultant against losses that result from errors or omissions in Client-provided information.

Client-provided information can be an asset or a problem – be sure that everyone's expectations regarding how it is to be used are clearly defined.

11. Indemnification – An indemnification is simply an agreement by one party to accept responsibility for someone else's loss. Common law provides that if a person's negligent act damages another person, the damaged party deserves compensation from the damaging party. In the practice of engineering in connection with indemnification, the operative term is "negligence". The indemnified party deserves compensation resulting from the negligent acts that occur during the performance of the Consultant's work. Expansion of this concept to include any or all

acts of the Consultant and collateral or consequential damages is inappropriate and should be avoided. Professional liability insurance provides coverage only for damages that result from the Consultant's negligent acts. The Consultant's professional liability insurance broker should review the Indemnification clause for insurability. Early review of this item when considering a contract is essential.

12. Change Management – Changes during the design almost always occur. It becomes imperative that how changes are to be managed be defined in the contract. Without this provision, the design process will likely be delayed while the parties agree on how to manage the change. The contract provision should address the form of the notice of a change and the time allowed for the party identifying the need for a change to notify the other party.

13. Purchase Orders – If the Client intends to use Purchase Orders to manage its procurement process, this needs to be addressed in the contract. The main problem for the Consultant with Purchase Orders is that they almost always contain a very broad Warranty section in the preprinted terms and conditions. The following is typical of Warranty provisions found in Purchase Orders:

"Supplier warrants that for a period of two years after the delivery of or performance of the Items, the Items will (a) be of merchantable quality; (b) be fit for the Company's specified purposes; (c) be of high quality, and be free from defects in material and workmanship; (d) comply with the most stringent of Company's or Supplier's specifications, performance guarantees and requirements; and (e) comply with all nationally recognized codes and established industry standards.".

Acceptance of the Warranty provisions can nullify the beneficial Standard of Care language that is so important in contracts for professional services. While requiring that the Consultant carry professional liability insurance, clients are frequently unaware that the Warranty provision effectively nullifies the professional liability insurance, which only covers the negligent acts of the Consultant. Further, Purchase Orders almost always contain very broad indemnification requirements, which may not be insurable.

It is essential that the contract clearly state that the pre-printed terms and conditions of the Purchase Order do not apply to the contract.

14. Subsurface Risks - A discussion regarding subsurface risks must begin with the definition of a surface. It can be any surface behind which the Consultant intends to make changes. It can be a retaining wall, a ceiling or wall within in a building, the ground surface, or any other surface.

    Certain investigative processes can determine, with some degree of certainty, the conditions that exist behind the surface in question. However, the cost and feasibility of determining all conditions behind the surface makes the results of the investigation an approximation at best. Accordingly, the process of coming to agreement with the Client as to how subsurface conditions will be determined and the risks attendant therewith becomes one of tradeoffs between the value and necessity of the more complete and accurate information and the cost of obtaining it.

    It becomes essential that the expected results of subsurface investigations are clearly defined, agreed to with the Client, and documented in the contract.

This may involve significant education of the Client.

The following are some thoughts about what should be included in the contract provision:

- Acknowledgement by the Client that it is not feasible and often not possible to determine all subsurface conditions that can affect the design.

- Identification of potential risks associated with not having complete subsurface information and agreement that those risks are acceptable to the Client.

- Acknowledgement by the Client that extrapolation of data between actual points that are investigated may not be a precise indication of what exists between the points.

- Acknowledgement by the Client that any investigation is valid only at the time of the investigation, as conditions may change between the investigation and actual construction.

- Agreement with the Client that a contingency fund will be included in the construction budget to accommodate differing subsurface conditions that only become known during construction.

The Consultant should carefully consider and chose the professional Subconsultants who work on their projects; particularly those are involved in performing geotechnical engineering investigations and evaluations. It is strongly recommended that the

Consultant utilize a Qualification Based Selection process for selection of all Subconsultants.

The Consultant and Client should acknowledge that below ground risks, which can affect construction cost estimates and Contractor claims due to changed conditions, include the accurate identification of the location of any rock and/or hard cemented soil layers that lie within the planned depth of excavations. The groundwater level at the time of construction along with likely fluctuations can also have a significant impact on the construction process and the potential for construction claims.

In order for the Geotechnical Engineering Subconsultant to provide a level of service that is normally expected by Consultants and Clients they should be provided with the following information prior to their development of the scope of work and budget for their services:

- Site map showing property boundaries and all known existing natural features and prior development on the site. Identification of any known hazardous materials on the site.

- Adjacent property owners and contact information if access on their property is required during the site investigation.

- Existing topographic data and proposed preliminary site grading plans.

- Details regarding the proposed site improvements, including dimension of all proposed structures, preliminary dead and live load conditions, required depths of excavation within and/or outside of the proposed structures, such as pits and basements.

It is almost impossible to avoid having subsurface conditions that differ from that which was expected. Be prepared for this situation by having a well thought out and documented change management procedure in place before design and construction begins.

15. Third Party Claims - Third party claims occur when a party other than the Client attempts to recover economic loss from the Consultant on whose design the third party has relied. On the surface, this does not seem reasonable or equitable. However, there is not consistency among the various States as to whether or not the Consultant has a duty or obligation to the third party.

Third party claims frequently come from subcontractors or vendors with whom neither the Client nor the Consultant has a contractual relationship.

The matter can be further complicated when a contractual relationship exists between the Client and say, a lender. This is particularly applicable when the Consultant's services include a study or analysis that may be of particular interest to the lender. The problem occurs when that which is of interest to the Client may be very different from that of the lender's. In which case, the Consultant's scope of services may not adequately address that which is of interest to the lender. All of this points in the direction of a need for a disclaimer in the contract between the Client and the Consultant stating that neither has any obligation or duty to any third party in connection with the services provided by the Consultant. In the case of the lender, it may be appropriate to consider entering into a co-client relationship (three-party contract) in which the needs of the Client and the lender are identified and mutual indemnifications of all parties are secured.

When there is resistance on the part of the Client to add the disclaimer to the contract due to not wanting to change a standard contract, adding the disclaimer to the scope of work may be a more acceptable alternative.

16. Stop Work Authority - Having the contractual right to direct a contractor to stop work should be held by the contractual entity that is in the best position to deal with the consequences of that action. In almost every case, that entity would be the Client. The moment that a "stop work" direction is received by the contractor, a whole host of complications and cost and schedule related claim items will begin to be tracked. Since dealing with these claims will ultimately become the responsibility of the Client, only the Client should make the call to "stop work". Unfortunately, not all Clients understand this and frequently attempt to shift that responsibility to the Consultant. Along with the contractual acceptance of the <u>authority</u> to stop the work, usually comes with the implied <u>responsibility</u> to stop the work when the Consultant deems that to be an appropriate action. Additionally, damages resulting from a Consultant stopping the work may not be covered by professional liability insurance. For example, when work is stopped, delays and inefficiencies almost always occur. These delays can result in consequential damages such as economic loss resulting from not having use of a facility. Consequential damages are almost always excluded from professional liability policies.

When asked by the Client to accept the contractual authority to stop work, here are some strategies for the Consultant to consider:

- Attempt to redirect the Client so that the client has the stop work authority.

- Offer an alternative that the Consultant will recommend that the work be stopped when, in the Consultant's judgment, it should be stopped.

- Offer an alternative that the Consultant will accept responsibility to reject or recommend rejection of work that, in the Consultant's opinion, does not meet the requirements of the contract documents.

- If the Client insists that the Consultant have the stop work authority, seek an indemnification from the Client for any and all damages that result from the Consultant giving stop work direction.

- When the Consultant has the stop work authority, the Consultant should also have a comprehensive role in the construction observation process.

- When the Consultant has the authority to stop work, extraordinary diligence in documentation of all construction activities is warranted.

Regardless of who has the stop work authority, it is essential that this authority be consistent in both the contracts between the Client and Consultant and between the Client and the Contractor. Additionally, stopping work should only occur with the Owners concurrence.

17. Curing a Breach – A breach of a contract occurs when one or both parties to a contract fail to perform in accordance with the terms of the contract or one party interferes with the performance of the other party. The breach can be deliberate or accidental and may be insignificant or material to the overall performance of the contract. The problem for the Consultant occurs when the contract does not address how a breach is to be addressed. In which case, the breach quickly jumps to a dispute that could have been avoided if an opportunity to cure the breach had been available.

    A typical contract provision that gives each party an opportunity to cure a breach simply requires that both parties notify the other party if they believe that a material breach has occurred. The notice usually takes the form of a communication identifying the breach and a warning of termination that will occur if the breach is not cured within a specific period of time. The notice must, of course, be consistent with the requirements of the termination provision of the contract. Once the notice has been given, good faith bargaining can begin to cure the breach and thus hopefully avoid a dispute or termination.

18. Dispute Resolution – As with managing change, agreement as to how disputes are to be resolved is necessary to have been established in the contract to avoid delays and further conflict while determining how the dispute will be resolved. The primary goal of the dispute resolution process is for the dispute to be resolved at the lowest level of stress, delay, and cost to all parties. Common methods of dispute resolution, listed in ascending order of difficulty and cost, are negotiation, mediation, arbitration, and litigation. Mediation and arbitration may be either

binding or non-binding. Due to the rules of discovery related to some arbitration, the costs are frequently very similar to those of litigation. However, damage to Client/Consultant relations is generally much higher with litigation.

19. Termination/Suspension of Consultant's Services – The termination provision in the contract should address the following:

- Which party can terminate/suspend

- Can termination/suspension occur with or without cause

- Can termination/suspension occur with or without penalty

- The amount of written notice required prior to termination/suspension

- How payment for services provided prior to termination is to be made

- How Consultant is to be paid for archiving of work prepared prior to termination/suspension

- That Consultant can suspend or terminate services due to non-payment by Client

<u>Expectation Management</u> is the ultimate goal of contracting for professional services so that when the project is completed, the Client and Consultant look at each other and say, "Yes, that is exactly what I expected". The following are some suggestions that can help achieve that goal:

1. Be as thorough as possible in preparation of the Scope of Work. Carefully identify what the

deliverables will be.

2. Document those potential scope items that are not included, especially those items that are discussed during scope negotiations and which the Client has specifically directed not be included.

3. Be sure that the Consultant's staff understands what is and what is not included in the Scope of Work.

4. Develop a close professional and personal relationship with the Client. Aim for a relationship in which the Client thinks of Consultant's staff as an extension of its staff.

5. Communicate often and well with the Client.

6. Document all communication with the Client. Send the documentation of what is understood to be the Client's direction back to the Client. That way, they can see in writing what the Consultant believes was communicated to it. That may or may not be what the Client intended but until they see it in writing, they only know what they intended to communicate to the Consultant. If the Consultant's understanding is incorrect, the Client's expectations will not be met. The Client will appreciate the opportunity to clarify the direction.

7. Address problems quickly. Unlike wine, problems do not improve with age, they only become worse. But, whenever possible attempt to have a recommendation for how to solve the problem before the problem is shared with the Client.

Expectation management is clearly the responsibility of the Consultant. The Consultant should accept the responsibility cheerfully and enjoy the benefits.

Scope of Work is the document that describes what the Consultant will prepare and deliver to the Client as its instruments of service, or documents. To the maximum extent practical, the Scope of Work (SOW) should be very detailed. The more clearly the SOW is presented, the higher the probability that the Client's expectations will be met. The following are suggestions regarding SOW preparation:

1. Provide a list of the titles of the likely sheets that are to be prepared.

2. Provide a list of the titles of the specification sections that are likely to be needed.

3. Provide a section entitled "Not Included" documenting those scope items for which agreement is reached during negotiations with the Client are not to be included. This is a critical element because neither the Client's nor the Consultant's Project Manager that negotiated the contract may ultimately be the individuals that direct the design process.

4. Consider a section entitled "Assumptions" that address things such as design standards that will be used, the timing of delivery of Client provided information, and other design-critical items.

5. Consider a section entitled "Contingency Items". This section would include items for which there is agreement that they may be needed but the need for which cannot be established before the design is undertaken. The advantage of having this section is that the SOW and budget may be agreed to beforehand so as to not delay the process when the need is determined.

6. Define, by Design Phase, what is to be provided as Basic Services and what is to be provided as

Additional Services. Basic Services generally include those tasks that can be clearly defined while negotiating the SOW. Additional Services are those tasks that may not be easily defined or whose required level of effort cannot be determined prior to execution of the design.

7. Consider including Construction Phase Services in the original SOW to avoid possible delays between the design and construction phases.

<u>Schedule</u> definition is essential to an orderly design process. For many projects the Client's schedule is the most important element of the contract and staying on the schedule takes precedence over almost everything else. The following are considerations regarding developing the contract Schedule:

1. Anticipated Notice to Proceed (NTP) date – Unfortunately, NTP on many projects does not occur when anticipated. This causes many problems for the Consultant who has scheduled staff and resources to begin on the project on a certain date. When NTP is delayed, the resulting cost and inefficiency to the Consultant may be recoverable from the Client, but only if the anticipated NTP date is documented in the contract.

2. Time allowed for each phase – For multiple phased projects, this time can be expressed in calendar days or working days. Calendar days are simpler as there is no question about how long that is. Working days must be accompanied by a definition such as "Monday through Friday except for Federal Holidays" or whatever definition can be agreed upon.

3. Define when work on the next phase can begin – Does the Client want the Consultant to keep working while the submittal is being reviewed (a "progress review")? Or, is the Consultant to stop work until comments are received and resolved (a "pens-down

review")? It can be either but it should be specified in the contract.

4. Define how long the Client has to review submittals.

5. Consider requiring written approval from Client after each phase prior to proceeding to next phase.

6. Seek agreement that delay caused by extended review will extend completion date.

7. Define overall completion date.

8. Provide some graphic presentation that makes understanding the schedule clear to all parties.

<u>Budget</u> is the term applied in reference to the amount of money available for payment to the Consultant.

1. The term "cost" when used in connection with a budget usually refers to the amount of money available to pay for labor (salaries, wages, and overhead related thereto) and other direct costs such as travel, reproduction, etc.

2. The term "labor," when expressed as a cost, can refer to only salaries and wages or a combination of salaries, wages and overhead related thereto.

3. "Overhead" is an expression of costs stated as a multiplier of salaries and wages that includes costs such as required and/or optional employee benefits, office rent and utilities, etc. The determination of the overhead multiplier is determined by the Consultant's accountant and certified to be true and correct. Many Clients require that the overhead rate prepared by the accountant be audited for compliance with a specific regulation such as the Federal Acquisition Regulations (FAR). Many Clients accept the audit

prepared for other clients without requiring a separate audit.

4. "Other Direct Costs" (ODC) are those non-labor related costs required to be expended in the performance of the design and include items such as travel, printing, permit fees, and shipping/mailing.

5. "Profit" is the term applied to the amount of money that the Consultant expects to be paid that is in addition to its labor, overhead, and ODC's.

<u>Basis of Payment</u> describes the method to be used to calculate the amount of money or fee owed to the Consultant. The following is a partial list of payment types in common use:

1. Lump Sum – This method of payment stipulates that for a specific SOW, the Client will pay the Consultant a specific amount of money.
2. Time and Materials (T&M) – The T&M method can frequently used when the scope and level of effort are difficult to establish or agreement between the Client and Consultant cannot be quickly reached. The element called Time includes the Consultant's labor, overhead, and profit with Materials including all non-labor related costs. Clients frequently limit the amount of all T&M costs.
3. Cost plus a Fixed Fee – In this method, the Consultant's compensation is based on its Cost (labor, overhead, and other direct costs) plus a Fixed Fee that, to some extent, represents the Consultant's profit. The Fixed Fee remains the same regardless of the costs. This allows an efficient Consultant whose costs are lower than anticipated to experience a higher level of profit when expressed as a percentage of the total fee earned. On the other hand, the Consultant's profitability is reduced when its costs are greater than anticipated. Clients frequently limit the amount of costs.

4. Cost Reimbursable – Not-to-Exceed - This method is frequently used in connection with Contingency Items in the SOW. Cost in this case is a combination of all labor, overhead, and profit expressed as a hourly labor rates plus materials and ODC's.
5. Percentage of Construction – This method determines the fee to be paid to the Consultant by simply multiplying an agreed upon percentage by the project construction cost. Of concern to the Consultant is how and when the construction cost is determined. If the construction cost is based on the low bid received, the unknowns associated with the bidding process make managing design costs to a budget extremely difficult if at all possible. The risk to the Consultant in this method can be very high. If, on the other hand, agreement as to the construction cost has already been determined as in a Construction Manager/General Contractor (CM/GC) delivery method, the risk of this method may be acceptable.

<u>Payment Terms</u> defines the timing of how money is paid to the Consultant. Payment may be made at the end of the Consultant's services or at specified intervals, commonly monthly, based on observable progress made during the invoice period.

<u>Contingent Fee Contracts</u> for professional design services are ones in which payment to the Consultant for his or her service is dependent on the actions of an entity other than the Client. A Client may attempt to tie payment of a Consultant's fee to receipt of a grant, obtaining a permit or a legislative funding process. A contingent fee contract likely creates the highest possible level of business risk to the Consultant. The risk is so high that entering into a contingent fee contract for many Consultants requires the specific approval from the highest levels of management. Contingent fee contracts are considered so problematic that many public agencies are prohibited by law from entering into a contract until the project is fully funded.

Generally speaking, consideration of accepting a contract containing a contingent fee clause should begin with refusal to accept the contract. Then move carefully away from refusal toward some justification for accepting it. The following are some things to take into consideration:

1. Is there is prior positive experience with the Client during which the Client has demonstrated correct judgment in evaluating the risk of the contingency being overcome?

2. Is there prior positive experience with the entity that controls the funding on which the fee is contingent?

3. Is the Consultant's economic situation such that the risk is warranted?

4. An alternative to refusal could be proceeding with negotiating the contract scope, schedule, and budget with the understanding that work will not begin until the contingency is removed.

5. Be aware that investing further time, energy, and money into the consideration process can create undue pressure to accept the contract containing the contingent fee provision.

6. Entering a contingent fee contract may be a violation of a Professional Society Code of Ethics or a State Engineering Registration Statute. Investigate this possibility before accepting a contingent fee contract.

All of the above considerations still contain considerable risk requiring managerial judgment and should be viewed as such.

Contract Signatory Requirements addresses the person signing the contract having authority to sign and thereby bind the entity entering the contract. Some Clients require a letter from the Corporate Secretary, bearing the corporate seal, stating that the person signing the contract has approval to bind the corporation for contracts up to a certain amount.

Notice to Proceed (NTP) is the Client's official notification to the Consultant that the design may begin. NTP is frequently given with the transmittal of the contract when executed by the Client. The NTP should include the specific date on which it is issued, thus establishing the date upon which future schedule milestones are determined. Many Clients will not issue NTP until they receive both an acceptable Insurance Certificate and an approved Quality Assurance/Quality Control Plan (QA/QC Plan).

Insurance Certificates are provided by the insurance broker to verify that the contract insurance requirements will be met. Contracts usually require that the Certificate state that the certificate holder (Client) be notified by a specified number of days prior to expiration or cancellation of the policy. Most insurance certificates are provided using a format developed by the Association for Cooperative Operations Research and Development (ACORD). The ACORD continuously reviews the changing regulatory requirements in all jurisdictions where ACORD forms are used, and revises forms as needed to keep them compliant. Prior to 2010, the ACORD form stated that the insurer would "endeavor" to provide the required notice. This language did not satisfy many Clients so the ACORD statement was revised to state that "notice will be delivered in accordance with the policy provisions". The new language revision was not particularly helpful either since the Client did not have convenient access to the "policy provision". The Consultant then has to review the policy provisions and provide them to the Client. Hopefully the policy provisions meet the contractual requirements. If they do not, the Consultant may seek agreement with the Client that the Consultant's providing the notice will be acceptable to the Client.

Contractor-Proposed Value Engineering occurs when by changing its materials, means, methods, or schedule, a contractor can offer "Value Engineering" cost savings that it will share with the Client. The Consultant of Record will likely be asked to review the proposed change. The risks associated with accepting contractor-proposed changes can be enormous especially compared with the compensation received by the Consultant to review the contractor's "Value Engineering" proposal. Some client's may expect the Consultant to review the contractor's "Value Engineering" proposal at no cost.

The following suggestions may be helpful in avoiding misunderstandings between the Client and the Consultant:

1. Include a provision in the contract between the Client and Consultant that calls for the Consultant's reviewing "Contractor Proposed Value Engineering" as an Additional Service.

2. Include a provision in the contract between the Client and the contractor requiring that the contractor provide certain specific information that will be needed by the Consultant to evaluate a contractor's "Value Engineering" proposal.

3. Include a provision in the contract between the Client and the Consultant that limits the Consultant's liability for review of contractor's "Value Engineering" proposals.

III. Implementing the Contract

A. Creating a culture of professional liability issues awareness

After the contract has been signed and Notice to Proceed (NTP) received, most Consultants give little thought to professional liability issues until there is a claim. Then they think about those issues a lot and wish that they had done so during the design process. To overcome this situation, Consultants can create a culture of professional liability issues awareness by providing regular and frequent reminders about these issues. This may be easily accomplished by including a brief professional liability issue topic in weekly planning meetings. Use of the various topics provided in this course at the meetings would be an efficient way to accomplish the weekly reminders. Additionally, use of these topics at the weekly meetings becomes a continuous training course for new employees.

B. Minimize Professional Liability Risk

There are many activities that the Consultant can use during the design that will both improve efficiency and may also help to minimize the probability of having a professional liability claim. The following are suggestions that have been found to be effective in this regard:

<u>Sharing the Scope of Work (SOW)</u> is the first step in managing expectations. The SOW is where the expectations are established. Every person on the design team should receive a copy of the SOW for the project or at least have access to it. While the Project Manager and some members of the team that were involved in developing and negotiating the SOW know what is included, others may be only partially

informed. They may be functioning largely on the basis of what they have heard or prior experience on similar projects.

Unless designers know what is included in the SOW and, more importantly, what is not, some will go beyond that which is in the SOW. Sadly, the Project Manager may only discover this when he/she observes that the project schedule or budget is in trouble. This situation is then compounded by the Project Manager having to decide whether or not to approach the Client with an awkward "after-the-fact" request for change in scope and/or additional compensation.

Verification that the design is being executed within the SOW is critical, especially early in the design process. A technique that can be helpful in this regard is to have an in-house 10% progress review during which the Project Manager can determine whether or not the design team is working within the SOW. This review can also avoid an unhappy schedule or quality surprise when it is time for the first formal submittal to the Client.

<u>Scope Change Identification and Documentation</u> must occur as soon as practicable. Changes in the SOW are common during the design process. Some are minor changes and some can have significant impacts on schedule and budget. Timely notification to the Client that the SOW needs to be revised is essential for maintaining good relationships with the Client and for the Consultant to properly manage the schedule and budget.

A Change in Scope form follows this section as Exhibit 1. It is simple, straightforward, and can be issued quickly by the Consultant to the Client providing the desired documentation as to why the Consultant believes that the scope needs to be changed. The form can also be used to notify the Client of the Consultant's estimate of potential changes in the schedule and budget. Use of the form can also establish with the Client that SOW changes are important, whether or not they require a change in schedule and/or budget. The form may also be used to document scope changes that occur early in the project that require minimal effort to implement so the Consultant may choose to make the change without schedule change or additional cost. In any event, the Client must be informed and agreement sought.

All too often Consultants accept verbal direction from the client that is not within the SOW, proceed with the change if is minor, believing that it can be absorbed within the budget. This sets an undesirable precedent with the Client who can come to expect that "scope creep" is an acceptable process. By using the form, future requests for change in SOW can be expected to be documented and, when appropriate, result in revisions to the schedule or budget.

But remember, the Change in Scope form is not a contract amendment, which is always required when the scope, schedule or budget changes. A sample Contract Amendment form follows this section as Exhibit 2.

(Insert Company Name or Logo Here)	
Change in Scope No.	
Client	Date
Contract or Spec Authorization Date: Engineer's Project No.　　　　　　　　　Client's Project No. Project Description:	
Nature of Changes:	Fee Adjustment: Previous Fee $ Increase/Decrease $ Revised Fee $ Estimate $　　　　Lump Sum $ Maximum Fee $
	Period of Service Adjustment: Previous deadline New deadline
Engineer's Project Manager _____　　Date _____	
Accepted _____　　Date _____ 　　　Client	

Exhibit 1

AMENDMENT NO. *[Insert Amendment No.]*

to

"*[Insert exact name of agreement or contract]*" dated *[Insert Date]*

PROJECT: *[Insert Project Name]*

CHANGE IN SCOPE, SCHEDULE, AND BUDGET

[Insert description of change in scope]

The additional budget for this Change in Scope is *[Insert Amount]* payable with the same terms as stated in the original Agreement.

All other terms and conditions of the original agreement remain in full force and effect.

If you agree with these provisions, please sign in the space provided below. Your signature also constitutes your Notice to *[Insert Consultant Company Name]* to Proceed with the work.

[Insert Consultant Company Name]   *[Insert Client Name]*

_____   _____

Name: Name:
Title: Title:
Date: Date:

Exhibit 2

Awareness of the Exclusions in the Professional Liability Policy is essential to prevent a Project Manager or design team member from inadvertently agreeing to or actually performing design activities that are not covered by the professional liability policy. Accordingly, every Project Manager should have a copy of the policy exclusions.

Activities that are typically excluded from professional liability policies include the following:

1. Claims based on the insured having provided express warranties, guarantees, assurances, etc. that are beyond the "standard of care"

2. Claims resulting from the manufacture or distribution of products

3. Claims based on personnel matters such as discrimination, harassment, wrongful dismissal, employee benefits, etc.

4. Claims based on construction performed by the insured

5. Claims based on damage to property owned, rented by, or under the control of the insured

6. Claims related to the design of nuclear energy facilities

There may be many other "exclusions" that are not mentioned above. Be sure to refer to the specific policy for the exclusions.

C. Improve Efficiency

<u>Planning for Meetings</u> has become an essential part of the design process. That which is accomplished at meetings and how meetings are documented is absolutely critical in managing everyone's expectations. The effectiveness of a meeting is largely dependent on how well the meeting is planned. Solid planning is essential, especially when the meeting is to be attended by multiple stakeholders and is to be held at a location several hours away from the Consultant's office where recovery from forgetting something can be very difficult. To manage this situation, using a checklist prepared well in advance of the meeting will be helpful. A sample checklist follows this section as Exhibit 3.

MEETING CHECKLIST

[Insert Meeting Date]

M = Meeting Date   M-XX = days prior to meeting   M+XX = days after meeting

1. Send request for meeting attendees to select a couple of dates within the next couple of weeks that they can attend the meeting. Suggest that the attendees visit the site before the meeting and attach any protocol required for site access. Request response within two days. M-14
2. Identify and begin preparation of material/exhibits needed for the meeting. M-12
3. Prepare draft agenda. M-12
4. Select meeting date based on availability of key attendees, compatibility with other activities, and availability of meeting location/conference room. M-12
5. Reserve meeting location/conference room. M-12
6. Reserve projector/recording device as appropriate. M-12
7. Notify attendees, using Outlook Scheduling, of time and place of meeting. M-12
8. Confirm meeting date, time and place with attendees – ask how many representatives will attend. M-10
9. Distribute materials to be reviewed prior to meeting. M-10
10. Send draft agenda to attendees for comment. M-10
11. Establish who will prepare/distribute meeting summary after meeting. M-10
12. Send directions to meeting location and parking suggestions to attendees. M-10
13. Hold in-house pre-meeting to clarify any questionable issues. M-5
14. Finalize meeting materials/exhibits. M-4
15. Prepare "start-of-meeting" announcements checklist. M-4
16. Prepare Meeting Note-taking Sheet. M-2
17. Finalize and reproduce Sign-In Sheet, materials/exhibits, and Agenda. M-1
18. Assign note taker for your organization. M-1
19. At beginning of meeting, confirm who will provide the meeting summary for review to verify that correct communication has occurred. M
20. At end of meeting, summarize action items, person responsible for action, and date by which action is due. M
21. Reproduce and distribute completed Sign-In Sheet at end of meeting. M
22. Send information agreed to during meeting to attendees. M+2 or as otherwise agreed.
23. Distribute draft meeting summary for review. TBD
24. Provide comments on draft meeting summary to preparer. TBD
25. Resolve comments and distribute finalized meeting summary. TBD

Exhibit 3

<u>Meeting Agenda</u> or the absence thereof, is the single most important determinant of the effectiveness of a meeting. The following are suggested guidelines regarding meeting agendas:

1. To improve the probability of accomplishing the purpose of a meeting, as a minimum, a printed agenda should be available to all attendees at the meeting.

2. A draft agenda should be routed to expected attendees for their input before the meeting. By so doing, the sense of ownership for the meeting outcome for all attendees is increased.

3. Input from other attendees helps the meeting organizer to be better prepared for otherwise unexpected topics that can come up during the meeting.

4. The finalized agenda should be sent to all attendees prior to the meeting so that they can also be prepared for the meeting.

5. The agenda lets the attendees know the order in which the items are to be covered so that the organizer's desired flow of the meeting is not interrupted by attendees unknowingly bringing up topics that are scheduled for later in the meeting.

6. In general, the agenda should be followed during the meeting; however, rigid adherence to the agenda is rarely necessary. Deviation from the agenda frequently occurs and can even be beneficial. However, if the meeting is getting out of control, the organizer can use the agenda to refocus the meeting.

A sample meeting agenda follows this section as Exhibit 4.

[Your Logo Here]	**[Your Company/Agency Name Here]**
	Agenda
	[Your street address here]
	[Your city, state, and zip code here]

MEETING DATE:	MEETING PURPOSE:
LOCATION:	

NEW BUSINESS:

OLD BUSINESS:

ACTION ITEM LIST

ITEM NO.	
1.	
2.	
3.	
4.	

Exhibit 4

Meeting Sign-In Sheet is the document used to capture the names and contact information of meeting attendees. - Having a clear record of who attended meetings is a vital part of the documentation process. Preparation of the Sign-In Sheet prior to the meeting accomplishes several benefits;

1. attendee presence will be documented

2. avoidance of the needless distraction of someone scrambling to start circulation of a blank piece of paper to record the attendees after the meeting is underway

3. establishes a desirable atmosphere that good preparation for the meeting has occurred

Towards the end of the meeting, the leader should offer an opportunity for anyone having not signed the Sign-In Sheet to do so.

At the end of the meeting, the Sign-In Sheet should be reproduced and distributed to all attendees so that they can immediately begin communicating with each other.

Those that attend telephonically should be identified as such on the Sign-In Sheet. For one-time meetings, a simple sign-in sheet works well. For meetings that are frequently attended by the same people, their name and contact information can be preprinted on the Sign-In Sheet, provided that a column next to each name is available for the attendee to initial indicating his/her attendance. A third type of Sign-In Sheet is for use when attendees are pre-registered, as with a seminar.

Sample Sign-In Sheets follows this section as Exhibits 5, 6, and 7.

[Your Logo Here]

Sign-In Sheet

Project Name: _____ Project Number: _____

Date: _____ Time: _____ ☐ AM ☐ PM Location: _____

Meeting Topic / Description: _____

	Name	Company / Agency	Phone Number	Email Address
1.				
2.				
3.				
4.				
5.				
6.				
7.				
8.				
9.				
10.				
11.				
12.				
13.				
14.				
15.				
16.				

Exhibit 5

	[Your Logo Here]					Sign-In Sheet
	Project Name:				Project Number	
	Date:		Time:	☐ AM ☐ PM	Location:	
	Meeting Topic / Description:					

	Name	Initials	Company / Agency	Phone Number	Email Address
1.	[Name]		[Company/Agency]	[(xxx) yyy-zzzz]	[abcde.fghi@jklom.com]
2.	[Name]		[Company/Agency]	[(xxx) yyy-zzzz]	[abcde.fghi@jklom.com]
3.	[Name]		[Company/Agency]	[(xxx) yyy-zzzz]	[abcde.fghi@jklom.com]
4.	[Name]		[Company/Agency]	[(xxx) yyy-zzzz]	[abcde.fghi@jklom.com]
5.	[Name]		[Company/Agency]	[(xxx) yyy-zzzz]	[abcde.fghi@jklom.com]
6.	[Name]		[Company/Agency]	[(xxx) yyy-zzzz]	[abcde.fghi@jklom.com]
7.					
8.					
9.					

Exhibit 6

[Insert Sponsor]
Sign-In

COURSE TITLE			COURSE CODE	INSTRUCTOR(S)	
LOCATION				DATE	TIME
Initials	First Name	Last Name	Agency or Company	Phone	Email
	[First]	[Last]	[Agency or Company]	[(xxx) yyy-zzzz]	[abc.def@hgijk.com]
	[First]	[Last]	[Agency or Company]	[(xxx) yyy-zzzz]	[abc.def@hgijk.com]
	[First]	[Last]	[Agency or Company]	[(xxx) yyy-zzzz]	[abc.def@hgijk.com]
	[First]	[Last]	[Agency or Company]	[(xxx) yyy-zzzz]	[abc.def@hgijk.com]
	[First]	[Last]	[Agency or Company]	[(xxx) yyy-zzzz]	[abc.def@hgijk.com]

Exhibit 7

<u>Meeting Summary/Minutes</u> document information communicated at meetings. This documentation may well be one of the most important elements in expectation management. Direction from the Client as well as decisions regarding various design elements are often made during meetings.

The most common forms of documentation are the Meeting Summary or Meeting Minutes. As the name implies, a Meeting Summary covers, in a topical way, the matters that were discussed and decisions reached, followed by further action that is required. Meeting Minutes are a more detailed coverage of the meeting in which the matters that were discussed are presented in the order in which they were discussed or re-discussed, followed by further action that is required. Unless greater detail is needed, as when claims or possible litigation are anticipated, generally, the Meeting Summary is used. The following are suggestions for effective meeting documentation

1. When the project manager is to lead a meeting, one or more other persons from that firm should attend the meeting. Prior to the meeting, the person responsible for taking notes should be identified. It should be determined beforehand whether a Meeting Summary or Meeting Minutes is to be prepared.

2. Consideration should be given to recording meetings, especially if Meeting Minutes are to be prepared. Relatively inexpensive digital recording devices that download directly to a computer are well suited for the recording. Attendees should be advised at the beginning of the meeting that the meeting will be recorded.

3. Scanning and attaching the Sign-In Sheet to the Meeting Summary/Minutes, and referencing it as

identifying the attendees is an efficient way to document who attended the meeting.

4. Draft Meeting Summary/Minutes should be routed to the attendees for comments/confirmation. A cutoff date for receipt of comments should be identified. Once comments have been resolved, finalized Meeting Summary/Minutes should be distributed to all attendees.

Be aware that although a Meeting Summary/Minutes can be good documentation of direction received that changes the scope, schedule or budget, the changes should be further documented in a Contract Amendment signed by both parties.

A sample Meeting Summary, which may be easily adapted to Meeting Minutes by changing the title, follows this section as Exhibit 8.

[Your Logo Here]	**[Your Company/Agency Name Here]** **Meeting Summary** *[Your Street Address Here]* *[Your City, State, and Zip Code Here]*	
ATTENDEES & DISTRIBUTION: *[Your Firm/Agency]*: *[Other Attendees]*:		MEETING DATE: LOCATION: PREPARED BY: ISSUE DATE:

MEETING PURPOSE:

SUMMARY:

ACTION ITEMS:

Action Item	Respondent
1.	
2.	
3.	
4.	
5.	

Exhibit 8

Permitting involves the Consultant assisting the Client in obtaining permission from a regulatory agency to construct a project. Of the three main causes of schedule disruption, underground utilities discovered late in the process, unexpected geotechnical conditions, and permitting, permitting is the only one that can be dealt with proactively. Prior planning can mitigate many of the problems associated with having permits in place when they are needed. Consideration of the following can expedite the process:

1. Avoid accepting responsibility for obtaining permits and licenses other than those required for the Consultant's practice. Permits are normally issued to Clients or construction contractors based on applications prepared by the Consultant or the contractor's engineer. The role of the Consultant should be to assist the permit holder in obtaining the permits, but should not include obtaining the permits, a process that is not under the control of the Consultant.

2. The level of effort needed in assisting the Client in obtaining permits is very difficult to estimate. Regulations governing permits frequently change during the design period, as do permitting agency staff that reviews and passes judgment on the permit application. Accordingly, permitting activities should not be included in basic services but rather in additional services for which compensation is based on some type of "time and materials" method.

3. Clients frequently want a "not to exceed" amount associated with permits. This requires that the specific permits that are included be identified in the "not to exceed" amount. Build adequate contingency into the estimating for the "not to exceed" amount.

4. The permitting process frequently requires meetings with the permitting agency. The number of meetings and the specific Consultant's staff that will be in attendance should be documented as being part of the "not to exceed" amount. If additional meetings or Consultant staff are required, an increase in the "not to exceed" amount is warranted.

5. During preparation and review of permits and even after permits are approved and issued, changes in permitting requirements, new permits, and changes in regulatory agency regulations or staff may occur. These changes should be specifically excluded from the scope of work and thus not included in the "not to exceed" amount. If these changes occur, the Consultant is entitled to a commensurate increase in the "not to exceed" amount.

6. Avoid agreeing to pay permit fees and later being reimbursed by the Client. Notify the Client of the exact name of the payee and amount and request that the Client provide the check. Plan ahead with plenty of notice to the Client for the check.

7. The Consultant's receiving its fee should not be contingent on the permits being obtained.

Assisting the Client in obtaining permits is almost always more difficult than can be anticipated in advance. Careful definition of scope and budget will help in controlling this process.

Documentation of the design and construction process is something that it is not terribly hard to do but it seems to be equally as easy not to do. In the rush of business, so many things seem self evident at the time and their documentation seems almost superfluous. But, memories fade, staffs change, and information becomes unavailable. An old Chinese proverb states, "The faintest of ink is more powerful than the strongest memory". Printing and posting the proverb near the telephone can be a helpful reminder to document telephone calls. Most Consultants will agree that no matter how good their documentation was, when the claim came, they wished that they had more documentation and that it was better! Additional incentive to document comes from trial attorneys who say that in professional liability cases, the litigant with the best documentation usually wins. The following sections address documentation of various parts of the design process:

1. Telephone Calls - Telephone call documentation begins with good note taking during the conversation. Many firms have a preprinted form for documenting telephone conversations – keep a supply of the forms close by. The person initiating the call should complete all the applicable information (date, time, person, subject, etc.) before starting the call. If the other party calls, just note the caller's name and time and fill in the rest later. Take notes while talking, using a short hand technique that can be filled in after the call. Use the initial of the caller followed by a brief statement of what was said in each statement. Use of a telephone headset can be helpful.

   Direction received by telephone from a Client can be conveniently documented by sending it in an e-mail to the design team members with a copy to the Client, requesting verification of the direction received. Seeing the written summary of what the Consultant understood the direction to be will alert the Client to any possible misunderstanding. If the direction requires a contract amendment, the Consultant should include that the amendment is being prepared.

   The method of archiving telephone conversation documentation will vary from one firm to another, depending on the firm's protocol for document control. However, it is not likely that keeping the documentation in more than one place or format will be criticized, especially if there is a claim or controversy. Retaining hand written notes taken during the conversation can add credence to the formal documentation.

2. Issue Tracking Log – An issue tracking log is a tool used to communicate issues that need to be resolved, identification of the party whose responsibility it to resolve the issue and the date by which action is required. While the need for the Issue Tracking Log is in direct proportion to the complexity and duration of a project, preparing and maintaining the Issue Tracking Log is beneficial for any project for the following reasons:

- Clients expect the Consultant to keep track of the details

- The Issue Tracking Log reduces the probability of the "I thought you were taking care of that" syndrome

- The Issue Tracking Log documents how design and construction issues are resolved

- Losing track of a detail during the design can trigger the "Omissions" part of an "Errors and Omissions" liability claim

- The Issues Tracking Log reduces the stress created by wondering if something has been forgotten

A sample Issues Tracking Log follows this section as Exhibit 9. The date that the form was prepared should be inserted in the "Insert Update Date." The Legend shows different type fonts that can be used to differentiate between whether the issue is a prior issue that has not been resolved (plain font), a new issue (bold font), or an issue that has been resolved and closed (screened font).

[Your Project Name] Issue Tracking Log
[Insert Update Date]

Issue Item No.	Issue	Originator	Date Issue Entered into Log	Assign to	Target Completion Date	Action	Resolution	Close out Date	Notes/Comments
1									
2									
3									
4									
5									
6									
7									
8									
9									
10									
11									
	Legend								
	Prior Open Issue								
	New Issue								
	Closed Issue								

Exhibit 9

3. <u>Site Photography</u> – Site photography documents the condition of the site before, during, and after construction and provides useful information to those involved in the design process. Insufficient documentation of site photography can greatly reduce its value. Many Consultants find a serious lack of or no documentation in project files regarding who took the pictures, when they were taken, and why they were taken. Some of these problems are eliminated with the advent of digital photography as the date can be conveniently imbedded in the data. GPS data can also be added in sophisticated camera systems. These options are set up within the software of the camera before taking pictures. Documentation of digital imaging can become a larger chore because there are generally more photos taken and the documentation can become overwhelming, leading to procrastination or not doing it at all. Immediate documentation should be a high priority.

Managing documentation of site photography may be improved by using the following suggestions:

- Plan the photo shoot before you leave the office. Prepare a customized Photo Log that you have readily available on a clip board as you take the pictures. The finalized photo log can be scanned and combined with the photos in the electronic Project File.

- Be sure that there is enough storage capacity in the camera – take more pictures than you think you will need. Edit out unnecessary photos soon after taking the pictures.

- Take some measuring device, such as a folding rule or yard stick that can be included in a picture for a scale reference.

- Take pictures of things like street signs or project entrance signs as references for the other pictures.

- Frame the pictures such that you get the most detail of the subject without a lot of irrelevant information around the important subject.

- On particularly wide scenes, take several pictures that can later be stitched together to make a panoramic scene. Use the horizon as a reference and keep it in the same vertical position on each frame. Select an object at the edge of one picture and be sure that the next picture contains the object.

- Download the pictures from the camera to a computer as quickly as possible and annotate the Photo Log while memories of the photo shoot are still fresh. Quickly transfer the pictures and the finalized photo log into the official electronic Project File.

A sample Photo Log follows this section as Exhibit 10. It was intended for a street improvement project but can be easily customized for any particular project.

Photo Log
[Insert Project Name]

Date								
Location								
Photographer								
Subject								
Job Number								
Photo Number	Intersection		Direction Looking	Comments/Reason for Taking Photograph				

Exhibit 10

Quality Assurance/Quality Control Plan (QA/QC Plan) is prepared by the Consultant and submitted to the Client for review. The Quality Assurance plan describes how the Consultant will verify that the Quality Control Plan is being followed. The Quality Control plan describes in detail the efforts that will be made by the Consultant to minimize the probability of errors or omissions occurring during design. Sample QA/QC plan templates are readily available on the Internet.

When having a QA/QC plan is a contractual requirement, it is recommended that contract language be included that allows the designer to deviate from the "plan" when appropriate in the designer's opinion. The following are some benefits of having a written quality control plan:

1. Improve the probability of the design being free from errors or omissions

2. Defending against an allegation that a designer was negligent for not having a quality control plan

3. Avoiding design deficiencies that have resulted in construction problems in the past

4. Preserving the firm's reputation for producing high quality designs

5. Differentiation during the selection process from other firms that do not have a quality control plan

The level of complexity of a quality control plan is dependent on the nature of the design. In some cases, a simple checklist is sufficient while other projects warrant a more comprehensive QA/QC plan. A running list of what caused construction problems should be maintained and added to the plan. This process can also take the form of "Lessons Learned" that should be documented at the end of each project and shared with all involved in future designs.

One last word of caution; having a QA/QC plan and not following it will likely be worse than not having a plan at all.

Certifications are frequently requested by Clients wherein the Client seeks statements from the Consultant regarding the quality or reliability of the services provided. Claims resulting from the use of the words such as assure, insure, warranty, guarantee, or certify are excluded from coverage under most professional liability policies. This is because they can be misrepresented as providing a guarantee or warranty. Accordingly, Consultants should avoid using these words whenever possible. Unfortunately, many Clients attempt to obtain "Certifications" from Consultants, not realizing that doing so may void the professional liability policy required by the contract. This matter became such an important issue in California that the professional societies had the following definition included in the Professional Engineers Act (Act):

California 2011 Professional Engineers Act, Paragraph 6735.5. Use of word "certify" or "certification"

The use of the word "certify" or "certification" by a registered professional engineer in the practice of professional engineering or land surveying constitutes an expression of professional opinion regarding those facts or findings which are the subject of the certification, and does not constitute a warranty or guarantee, either expressed or implied.

If the Consultant's state does not have a similar definition in the Act that regulates its practice, Consultants should work through their professional societies to have it added.

When faced with a situation in which the Consultant is asked to provide a certification, the following are some suggestions regarding how to deal with it:

- Discuss the implications of providing the certification with the Client and attempt to have the requirement removed

- Modify the certification form to use less onerous words such as "the design professional states that"

- Add the above definition, adapted to Consultant's practice, to the form

Constructability Reviews are performed by a construction specialist that has actually constructed a project like or similar to the one that is being designed. Designers, often with minimal construction experience, conceive designs that seem appropriate but overlook significant elements that may not be constructible. These oversights would be readily apparent to an experienced construction specialist and are the things of which problems during bidding or claims during construction are made. These can be easily overcome with a constructability review, the earlier in the design the better, but certainly before the project is advertised for bidding. Projects that are put out to bid with constructability issues will result at best, in questions during bidding requiring addenda or at worst, claims during construction.

The following considerations may be helpful regarding constructability reviews:

- Construction contractors are frequently willing to review a design and make comments as to the constructability. Many will provide the review without cost or obligation.

- It is best to seek a review from a contractor that does not intend to bid on the particular project. Otherwise, other bidders may protest that the reviewing contractor had an unfair advantage.

- Construction Management firms frequently offer this service or may be willing to provide it if requested. The fee charged for this review is money well spent.

Construction Contingency Fund is established to facilitate management of costs related to changes that occur during construction. Rarely are projects constructed without any change, almost all of which involve an increase in the contract price. Experienced and technically cognizant Clients recognize this and establish a contingency fund right from the beginning. They accept that neither they nor the Consultant are perfect and therefore will not be able to foresee everything that will occur during construction. However, less experienced and some more difficult Clients may be unaware of the need for or unwilling to accept this premise. When changes or additions are found to be needed for whatever reason, the latter group is prone to allege that a design deficiency has occurred and look to the Consultant to pay for the change or addition, triggering a professional liability claim. The following are considerations regarding having a construction contingency fund:

- The existence of a construction contingency fund is an acknowledgment that the Client agrees that there will likely be a need for one. Without one, the relatively simple activity of accomplishing a change or addition to the project will be made significantly more complex by having to deal with identifying a funding source.

- In the absence of a construction contingency fund, some Client project managers are resistant to presenting a request for additional funding to their political board or investors. They fear being viewed as weak or inadequate for not having anticipated the possible need for the change or addition. To overcome this problem, they may resort to looking to the proceeds of a professional liability claim as the funding source. This can begin a time, energy, and money consuming activity that could have been avoided.

- The management level at which access to the construction contingency fund can be made should be at the lowest level consistent with the complexity of the project. Frequently there is a monetary limit above which access to the contingency fund must have the approval of the next higher Client management level.

- Similar reasoning can be applied to having a contingency fund related to the design process.

Early educational effort with Client regarding the need for a contingency fund can pay big dividends during execution of the design or during construction.

<u>Accounts Receivable</u> management is critical to the Consultant's financial wellbeing. Sometimes, when a Client isn't paying its invoices on time, it can be "the canary in the coal mine". While most Clients start out with good intentions of paying invoices in accordance with the terms of the contract, if they get behind, they can be tempted to seek relief by alleging a design deficiency and perhaps followed by a professional liability claim. The following are some proactive actions that can be taken in an attempt to discourage potentially financially damaging Client behavior:

- Call the Client as soon as the invoice has had time to be received. Ask if they received the invoice.

- Ask if they understand the invoice and if it is in an acceptable format with appropriate back-up.

- Ask if they agree that the value of the services rendered during the invoice period is consistent with the amount of the invoice.

- If the answer to all of the questions is "yes", ask when receipt of payment can be expected. If the answer is "no", the Consultant's project manager must work with the Client to get to "yes."

- Document the phone call with an e-mail to the Client with a copy to the project manager.

- Follow up immediately if the payment doesn't arrive on time.

The person making the call must have excellent diplomatic customer service skills to avoid offending the Client. Use of the above procedure can reduce payment time by as much as two-thirds. Additionally, it sends a message to the Client that on time payment is important. If the Client is having a cash flow problem, the Consultant may receive preferential treatment when the Client is deciding which invoice they should pay, believing that a slow-pay situation will not be allowed to slide.

When the above has not been effective, immediate attention is needed to remedy the situation. A reasonable approach could be for the Consultant to let the Client know that resolving the matter is desired before proceeding with legal remedies available under the contract. If the Client convinces the Consultant that payment is forthcoming, but not right away, the Consultant should ask for a time frame over which the Client expects to be able to make the payment. If that period is acceptable, the Consultant may wish to use of a promissory note that includes the following:

- A "whereas" statement in which the Client agrees that the services have been properly rendered and that the total amount is due and payable.

- A repayment schedule that stipulates the payment amount due for each and columns that show the principal amount due, interest on the principal at the amount allowed in the contract, and total payment due.

- A signature block for the client to sign and date the note.

The promissory note redirects the process back into conformity with the terms of the contract and away from a dispute situation. Of equal importance, the Consultant will then have a document, signed by the Client that preempts any future claim that payment is being withheld due to design deficiencies or dissatisfaction with the Consultant's services.

If the Client misses the first payment called for in the promissory note and the Consultant still prefers not to pursue legal remedies, the use of post-dated checks may be an option. This option usually is used when the Consultant believes that the Client has some assets but is choosing to give other obligations preference. This process involves obtaining post-dated checks from the Client, one for each month in accordance with an amended repayment schedule in a new promissory note. Because the Client knows that the Consultant is going to cash the check on the date it is due, that payment moves to the top of how payment obligations are to be prioritized. If the Client stops payment on the post-dated checks, the Consultant's claim becomes even stronger if litigation is pursued.

Having to go to extraordinary measures to get paid is a painful waste of time and energy. The following suggestions may be helpful in this regard:

- Choose Clients carefully. If getting paid by a Client has been a problem in the past, shame on the Client. If it happens again, shame on the Consultant.

- Obtain a retainer that is held and applied to the final, not an intermediate, invoice.

- Have a provision in the contract that allows suspension/termination of services for non-payment of invoices.

- Pursue overdue receivables immediately.

Shop Drawing Management is a crucial element of the design process that begins well before bids are received. Agreement between the Client and Consultant is essential regarding the role of shop drawings in the construction process, the Consultant's role, and how the review of shop drawings is to be managed. The following address some of the considerations in this regard:

- Reach agreement with the Client regarding the purpose and limitations of shop drawing review by the Consultant.

- Reach agreement with the Client that clearly defines the process that will be used for the review of shop drawings, especially the time allowed for shop drawing review.

- Identify those construction elements for which shop drawings are to be submitted and which are not. Exclude review of the Contractor's means, methods, sequences, or temporary construction items.

- Obtain agreement from Client regarding the content of the Shop Drawing Review Stamp that is to be used.

- Require in the specifications that the Contractor review the shop drawing prior to submittal and represent that the item meets the requirements of the contract documents. Require that the Contractor call to the attention of the reviewer those elements in the shop drawing submittals that do not comply with the requirements of the contract documents.

- Require in the specifications that a Client-approved Shop Drawing Submittal Schedule be in place before Notice to Proceed is given. If the schedule contains anomalies that can diminish the effectiveness of the review, require that the schedule be revised and resubmitted.

- Require in the specifications that shop drawing submittals be numbered sequentially to minimize the probability of having shop drawing misplaced and unaccounted for.

D. Liability

Liability as addressed herein relates to professional liability and does not include "strict liability", which may apply to some design professionals. Liability occurs when the negligent act(s) of an individual or company results in damage or injury to someone else. Two essential elements necessary for liability to exist are:
    1. Damages
    2. Negligence

Both elements must be present for liability to exist. Often when a liability claim is made, this principle is ignored or one or both of the elements may be overstated. This may be in attempt to intimidate the entity accused of alleged liability and obtain an offer of settlement without having to prove that either element exists. This, of course, can be easily overcome by a competent attorney. Receiving a claim of alleged liability can be a frightening event because just responding to the claim can consume significant time, energy, and money, none of which is likely to be compensable. Care should be taken when making the initial response to a professional liability claim not to overreact. Obtain the facts needed to defend against the claim first and then develop a structured defense based on input from those skilled in the applicable technical and legal areas.

"Damages", in the context of professional liability, is the actual additional cost to the Client resulting from a negligent error or omission. Of the two required elements of liability, "damages" and negligence, "damages" is most likely the more objective. However, arriving at a precise amount of "damages" is frequently accomplished by negotiations between the parties.

The following are some considerations regarding damages:

- Rework – if previously completed construction is found to be incorrect due to a negligent error or omission and has to be removed and replaced, the cost of this process will almost always fall into the category of damages.

- Work delays and inefficiencies – when tied to the damages, these costs are generally accepted as part of the overall damages cost.

- "Betterment" – Damages does not include a "betterment", which is construction not included in design or contract amount. If the Client has not previously been contractually obligated to pay for an omission that is later found to be needed, that cost should become the obligation of the Client. An example would be a driveway turnout from a street that was being constructed that was requested by a property owner after the design was completed.

- Premium Costs – an amount that is normally determined by taking the difference between what is estimated to be the cost of adding an omission by change order and what is estimated to be what would have been the cost of the omitted item had it been included in the original bid. This cost is normally added to other elements of damages.

Negligence, in the context of professional liability, is the failure to perform within the Standard of Care that is defined in the contract. Of the two required elements of liability, damages and negligence, negligence is most likely the more subjective. It is also the element of liability over which the Consultant has the most control.

Consideration of whether or not negligence has occurred can become a matter of opinion as expressed by other design professionals. Three scenarios come to mind for how negligence may be determined:

1. If there is a contractual obligation that established the entity that determines the presence of negligence, the matter is referred to that entity for determination.

2. If there is a contractual obligation to resolve a dispute or professional liability claim by arbitration, an arbitrator is agreed to by both parties and he or she makes the determination. Expert witnesses may or may not be called to testify regarding their opinion of the presence of negligence. The arbitrator decides whether or not negligence was present. The arbitrator's finding may or may not be binding depending how the contract addresses arbitration.

3. If a professional liability claim proceeds to litigation, expert witnesses for both the plaintiff and the defendant will express their opinion regarding whether the defendant was negligent. The jury will then be called on to decide which witnesses were more credible, establishing or rejecting the presence of negligence.

<u>Negligence Avoidance</u> begins with the knowledge of what activities have frequently resulted in negligence in the past. These activities can be divided into two broad categories of how negligence is normally evaluated; objective and subjective. Objectively viewed negligence involves situations in which the Consultant either did or did not do that which was clearly expected of someone operating within the "Standard of Care". Subjectively viewed negligence requires judgment and opinion on the part of the viewer.

The following are some examples of negligent design activities:

Objective

- Design not in accordance with the latest published Client's Standards, Manuals, Guides, and Forms

- Industry Standard of Practice not met (not just Client's standards)

- Design process not in compliance with Consultant's published Quality Control Procedures

Subjective

- Technical competence of consultant (junior designer, when senior designer required, generalist when specialist required....)

- Site conditions not appropriately considered in design

- Lack of appropriate care and professional judgment

The above list is clearly not exhaustive. The Consultant should identify other situations that have or will likely result in negligence and share them with its staff.

Professional Liability Claims Process defines how a Client handles alleged Consultant errors and omissions (E&O) in its designs. Many public agency Clients have published procedures that define the process but many Clients do not. When a Client indicates that it is going to file a professional liability claim, the Consultant should ask the Client for a copy of its professional liability claims process. If the Client replies that it does not have one, the Consultant may have an opportunity to provide some guidance regarding how to proceed managing the claim. This may inure to both the Client's and Consultant's benefit since both should want the claim to be dealt with quickly and fairly. The process used by a local public agency may be a good place to start.

The following is a list of the common steps frequently found in a public agency published process:

- Objective – Work together to minimize effects of E&O Issue (Problem)

- Client provides oral and written notice of basis of claim and requests that Consultant provides specific documentation

- Consultant provides Client-requested documents

- If Consultant believes it is entitled to additional compensation to provide alleged omissions, Consultant provides basis of justification for additional compensation within 7 days

- Client Contract Administrator (CA) and Consultant meet to discuss how to mitigate the problem

- If unable to resolve the problem, CA requests that individual identified in the contract that determines whether a Consultant's performance is within the Standard of Care (usually the Client's Chief Engineer) initiates an internal review

- Chief Engineer determines whether or not the Standard of Care has been met

- If Chief Engineer determines that Standard of Care was met, the claim is withdrawn

- If the Chief Engineer determines that the Standard of Care was not met, Client pursues Cost Evaluation and Recovery
- If Consultant agrees that Standard of Care was met, negotiations begin to settle the claim
- If Consultant believes that the Standard of Care was met, dispute resolution provisions of the contract are applied

IV. During Bidding and Construction

A. Before Construction Begins

When the design has been submitted and approved by the Client, usually the next step is for the Client to receive bids for construction of the project. The Client may choose to manage the bidding process with its own staff or it may request assistance from the Consultant. When the Consultant's Scope of Work includes responsibility for administering the process of receiving bids, care must be taken to avoid having one or more bidders protest the results of the process. A bid protest may be filed when a bidder claims that another bidder had an unfair advantage. This claim is usually based on an allegation that certain information was not available to all bidders at the same time.

Some suggestions for minimizing the probability of a bid protest follow:

- In the Instructions to Bidders, identify the person to whom questions during bidding are to be asked. Include contact information – FAX No. and e-mail address. State that no questions will be answered verbally. Specify a cut-off date for questions.

- When providing written response to questions, repeat the question exactly as submitted without any editing.

- Provide written answers to questions to all bidders at the same time. This is usually accomplished by posting the questions and answers on the Client's website.

- Answers that revise quantities, schedule, specifications, drawings, or bid date usually require an Addendum. State, in the answer to questions, whether an Addendum will be issued in connection with the question.

- Allow enough time for bidders to adjust bids based on answered questions. If questions are received very late in the time allowed for questions, consider issuing an addendum revising the bid date to provide additional time.

- Issue a new bid form with each addendum including acknowledgment of having received all addenda.

- Review and obtain approval of the above process with the Client prior to advertising for bids.

Contractors frequently wait until very late in the time allowed for questions to avoid exposing issues to other contractors, thus limiting the time that the other contractors have to react to the issue. The Consultant should reserve time in its calendar for answering bidder's questions, most of which will occur just before the bid date.

This process may seem excessive and tedious, depending on the complexity or size of the project and common practices of your Client. However, having to rebid a project is a significant waste of everyone's time, energy, and money. If the reason for having to rebid a project can be traced to inappropriate Consultant action, the Client may suggest that the Consultant reimburse the Client for the cost of rebidding.

B. During Construction

Request for Information (RFI) is the name given to the process used by the contractor to obtain clarification of various elements of the design documents. Receiving an answer quickly is critical to the contractor's maintaining the contract schedule. Excessive review time will likely result in a contractor filing a claim for delay. Agreement should be obtained between the Client and the Consultant as to what is the allowable time for responding to RFI's.

RFI's should be numbered by the contractor and the Consultant should maintain a log that documents the RFI process. A sample RFI log follows this section as Exhibit 11.

Page _____

RFI Log

RFI Number	Date Rec'd	Issue	Assigned to	Date Closed

Exhibit 11

Shop Drawing Review involves verification that materials proposed by the contractor comply with the contract documents. Clients expect that the Consultant is the most qualified to evaluate whether or not the specific construction items being offered by the contractor are in compliance with the contract documents. Accordingly, this responsibility is almost always included in the Consultant's Scope of Work. Here are some things to keep in mind as shop drawings are reviewed:

- The contractor is the party that is responsible for providing construction items that are in compliance with the construction documents.

- Avoid accepting any responsibility for the contractor's means, methods, and schedule. This includes being involved in reviewing shop drawings for how the contractor will fit the various elements together.

- Establish a formal administrative process for reviewing shop drawings that includes logging, tracking and follow-up. Share it with the contractor so that they know what to expect.

- Require that the contractor provide sequential numbering of all shop drawings. This can avoid a submittal being misplaced and no one knowing that it is not being reviewed.

- Require that the contractor stamp the submittal indicating that it has been reviewed and approved by the contractor and that it complies with the requirements of the contract documents. Insist that the contractor identify any item being offered that does not meet the specification. If by inspection, it is apparent that an appropriate review by the contractor has not occurred, return it to the contractor without review. Do not accept or review any shop drawings submitted directly by subcontractors or material vendors.

- Avoid accepting a contractor's request to deviate from the requirements of the contract documents to accommodate the contractor's means, methods, equipment, or schedule. If unavoidable, any such deviation should be made through the change order process and be agreed to by the Owner.

- Only review those shop drawings that are listed in the contract documents as being those that are to be reviewed. If you receive other shop drawings, return them indicating that they have not been reviewed.

- Carefully select the reviewer that has the right skill level. Avoid the temptation to select the reviewer on the basis of availability or any other lesser criteria.

- Complete the shop drawing stamp in accordance with the process agreed to with the Client.

- Complete and return the processed shop drawing within the specified time period. Both the Client and the Consultant have responsibilities in this process. Both must avoid any delay in processing of shop drawings by giving this activity the highest priority.

Shop drawing review is a critical element in the construction process. It should be taken very seriously – the consequences of not doing so can be catastrophic, resulting in costly damage and loss of life.

A sample shop drawing review follows this section as Exhibit 12.

Page _____

Shop Drawing Review Log

Shop Drawing Number	Date Rec'd	Item Description/Specification Section	Assigned to	Approval Status	Date Returned

Exhibit 12

<u>Performance Bonds</u> are surety bonds whose purpose is to provide Clients protection from financial loss occasioned by a contractor not completing a project in accordance with the contract documents.

Calling on a bond company to complete a project may occur when a contractor files for bankruptcy or otherwise refuses to complete the project. When this occurs, many bond companies will complete the project or provide funding to complete the project with minimal resistance. However, when the Client elects to terminate the contract as a result of, in the Client's opinion, the contractor's poor performance or breach of the contract, an entirely different situation occurs, which will likely be met with resistance from the bond company. It may become necessary for the Client to sue the bond company to complete the project or gain access to the funds.

The Consultant should take the following into consideration both before and during construction:

- Avoid giving the Client advice about the type and amount of coverage needed or criteria for selection of a bond company. Rather, encourage the Client to retain a risk management consultant for these services. Some Clients that do not wish to undertake that process simply require that the bond company be on a list of companies that are acceptable to a specific public agency.

- Define in very clear and specific terms in the General Conditions those situations that justify declaring the contractor in default.

- Define the time limit for the contractor/bond company to cure the problem and thereby avoid termination.

- Notify the bond company as soon as it becomes apparent that the contractor may be having difficulty avoiding a default situation.

- Recommend that the Client's attorney be informed of the situation.

- Recommend that, when the contractor's performance is poor, the Client withhold payment to the contractor such that if it becomes necessary to terminate the contract, the Client will have sufficient funds to complete the project.

- Give the required notice to the contractor/bond company of the pending termination, advising that the termination may be averted if the deficiency is cured within the specified time limit.

- Follow the requirements of the General Conditions very carefully.

Thorough documentation of any action involving a performance bond is essential.

Record Documents are the contract documents in which changes made by the contractor during construction are documented. Frequently, and incorrectly, record documents are referred to in Client contracts as "as-built" documents. Record documents can be a significant source of unmet Client's and other's expectations. The term "as-built" implies that the information provided is a complete and totally accurate representation of what was actually built, which it rarely is. To minimize this problem, the term "as-built" should not be used in the Consultant's contract as being something for which the Consultant is responsible. When the Scope of Work includes preparation of record documents, consider the following:

- If the Client does not ask for record drawings, the Consultant should not offer to provide them.

- Explain to the Client how record drawings are prepared, that they will not be verified by the Consultant, and the limitations of use thereof.

- Consider shifting the responsibility for preparation of the record documents to the contractor.

- Include providing the changes made by the contractor during construction in the "Information Provided by Others" section in the contract with the Client.

- Include an explanation of record documents in the "Definitions" section of the contract with the Client.

- Require in the construction contract documents that the contractor maintains a contemporaneously prepared documentation of changes that occur during construction.

- If the Consultant's Scope of Work includes construction observation, ask to see the contract document set being used by the contractor to record changes each time the construction site is visited. If none are available, or there is no indication that the changes are being documented, notify the Client and request assistance in getting the contractor to maintain the records.

- Stamp each document with a warning regarding the reliability of the information thereon.

Record documents are of great value to the Client and participation in creating them frequently falls to the Consultant. The Consultant should make sure that the Client understands the Consultant's role, responsibilities, and limitation in connection with record documents.

V. Summary and Recommendations

A. Summary

Now that a better understanding of the topics addressed in this book has been absorbed, the project managers are ready to more efficiently manage the issues that they face every day. More time should be available for managing the design process so as to improve the quality of construction contract documents and management of the construction process.

While universal acceptance by project managers is not anticipated, many would agree that of the approximately fifty topics addressed in this book, the following list of "top ten" considerations are the most critical:

10. Accounts Receivable – When a Client falls behind in paying, this can be the precursor of other problems. It frequently can mean that they are running out of money or are unhappy with the Consultants services. Either way, immediate action is called for. Try to get at what the problem is and get it fixed.

9. Purchase Orders – The Standard Terms and Conditions (fine print) that come with purchase orders are not for the Consultants benefit. Accepting them can void professional liability insurance coverage.

8. Create a Culture of Awareness to Professional Liability Issues – Regular reminders of the topics addressed in this book will be beneficial in this regard.

7. Contractor Requested Changes from Contract Document Requirements – These changes are almost always requested to benefit the contractor. Acceptance of these changes rarely inures to the benefit of the design professional or the Owner.

6. Stop Work Authority – Stopping a project almost always initiates a host of complex, expensive, and time consuming activities that frequently have professional liability and legal consequences. Stopping work should only be done with the concurrence of the Owner.

5. Shop Drawing Review – Proper shop drawing review is extremely important. Inadequate shop drawing review can have consequences ranging from increased cost to loss of life.

4. Scope of Work – It is only with a clear scope of work that the expectations of all parties can be expected.

3. Expectation Management – Many Clients will expect perfection from the Consultant. Unless this expectation is properly managed from start to finish, having a successful project is highly unlikely.

2. Constructability Review – There will always be a constructability review. The best time to have one is at the preliminary design phase but always before advertising for bids. Otherwise, the review will occur during bidding or construction with potentially costly consequences.

1. Documentation – No matter how much documentation is prepared, when problems arise, most project managers will agree that they wished that there had been more documentation and that it had been better.

B. Recommendations

The next step should be to share this information with other professional staff members so as to multiply the benefits of what has been learned. Make a commitment to mentor recent graduates and others by setting up a regular training schedule. Informal luncheons, where the professional staff sits around eating lunch while they discuss one or more of these topics, seem to work well.

Made in the USA
Columbia, SC
20 June 2017